Declassified Nerve Gas Production Processes

GB, VX, and BZ

by Yogi Shan

Completely revised 2nd Edition

Version 3.0

ISBN-13: 978-1-5116-8967-0 (Printed Paperback)
ISBN-10: 978-1-5116-8967-6 (Kindle e-version "Soft Copy")

Printed by CreateSpace.com
Distributed by: CreateSpace.com, amazon.com, amazon.ca, amazon.co.uk, and many others

Comments, inquiries, insults, and ordering information to: yogishan2000@yahoo.ca

Table of Contents

G-Agents

V-Agents

The Secret of "Mr. Death": A Homemade V-agent

BZ

Introduction

ATTENTION!

PLEASE NOTE:

1) Intent or attempting to manufacture nerve gas/nerve agents is highly illegal in the U.S. under Federal law, not to mention incredibly dangerous and likely to be lethal/fatal.

2) While the syntheses described are technically possible for a trained and well-equipped organic chemist, they pose insurmountable difficulties for the amateur, not to mention the difficulties in obtaining the precursor chemicals, organic chemistry glassware, powerful fume hoods, protective masks and specialized clothing, oxygen supply, and auxiliary equipment (partially thanks to the people manufacturing illegal drugs).

3) The agent manufacturing processes described require special techniques and equipment for the handling and storage of dangerously toxic (poisonous) substances. The paper (article) in the "Canadian Journal of Chemistry" ("Can. J. Chem.") Volume 38, pages 1416-1427 (1960) contains excellent illustrations of the typical chemical glass-ware required for manufacture in a laboratory setting.

*   *   *   *

U.S. Patents are available on the Internet for free. ("freepatentsonline.com" is the best source.) They are also available free (in a less user-friendly form) from the U.S. Patent Office (uspto.gov).

The British VX patents were withdrawn after a publicity kerfuffle in the 1970's, but are available from the U.S. Patent Office by regular mail. Detailed summaries of these patents – and any chemical patents – are also available in the journal "Chemical Abstracts", available in any university science library.

Though "nerve gas" is the term in common usage, "nerve agent" is the more proper military/technical/scientific term.

*   *   *   *

Lethal Doses:

GB/Sarin: 7.5 mg

VX: VX when inhaled is 7 times more toxic than GB (i.e, 1.1 mg);
10 mg   (through the skin);
.00045 mg (Subcutaneous).

Dr. MacArthur: "[A] typical nerve agent – I am talking right now about GB – requires one ton of agent dispersed in the air to produce 50% casualties to unprotected personnel over an area of about 1 square mile."

*   *   *   *

GB Stabilizers (To prevent decomposition during long-term storage):

5% tributylamine, or 2-4% by weight dicyclohexylcarbodiimide, or with diisopropylcarbodiimide

(DICDI) [U.S.P. #4,012,464 (1977), filed in 1965; i.e. classified for 12 years]

*   *   *   *

General Nerve Agent Chemical Structures:

```
    O              S                  O
    ||             ||                 ||        R'2
  R-P-F          R-P-F             R- P-S-Et-N<
    |              |                  |         R'2
   OR'            OR'                OR"

 G-agents      Thiosarin            V-agents
```

Specific Nerve Agent Chemical Structures:

```
        O                        O                          O
        II                       II                         II
Me – P – S-C-C-N (iPr)2   Me – P – S-C-C-N (Et)2   EtO – P – S-C-C-N (Et)2
        |                        |                          |
       OEt                      OEt                        OEt

        VX                       VM                    VG (amiton)
```

```
        O                        O
        II                       II
Me – P – S – C – C – N (Et)2   Et-- P – S – C – C – N (Et)2
        |                        |
     O-C-C-Me2                   OEt

        VR                       VE
```

U.S. Nerve Agent Precursor Code Names

Methyl phosphonous dichloride (MePCl2): SW
Methyl phosphonic dichloride ("dichlor"/(MeP(O)Cl2): DC
Diethyl methylphoshonothionate: TRS
O,2-Diethyl Methylphosphonothiolate: OSDMP
Trimethyl phosphite: TMP
Diethyl Methyl Phosphite: MP
PCl3: TH
Dimethyl Methylphosphonate: DMMP

G-agent and V-agent Thickeners

G-agents in particular, are thickened to reduce their volatility, thus increasing their persistence and effectiveness.

0.1-2% polymethyl methacrylate (also known by the trade-names Plexiglass, Lucite, and Perspex) is used for G-agents, and 0.1-2% polystyrene (or Plexiglass) is used for V-agents. [U.S.P. #3,868,446 (1975), filed 1963] [U.S.P. #4,708,869 (1987), filed 1969)]

G-Agents

The DMHP Process

The DMHP (Dimethyl Hydrogen Phosphite) Process was the first process discovered to synthesize "nerve gas", specifically Sarin (later known as GB). It was invented and produced in multi-ton quantities (but never used) by the Nazi's in 1938, just before World War 2, and appropriated by both the Americans and the Russians, as they both made a concerted effort to pick over the rubble of the defeated post-War Germany in 1945.

The DMHP Process is a five-step process for the preparation of GB via methyl phosphonic dichloride (known as "dichlor"). It is not considered that great a method, so U.S. post-WW2 research developed many other alternate syntheses (described later) in their search for the "perfect synthesis."

Step 1:   PCl3 + 3MeOH + 2NH3 --> (MeO)2POH + MeCl + 2NH4

The reaction – which produces DMHP – is performed in a hydrocarbon solvent, like toluene.

                                    heat
Step 2:   (MeO)2POH --> [MeP(O)(OH)]2O + Me2O + (MeO)(OH)P(O)Me
                                    "pyro"

The mixture of the three products is known as "pyromix".

Step 3:   [MeP(O)(OH)]2O + 2PCl3 + 2Cl2 --> MeP(O)Cl2 + 2POCl3 + MeCl + 2HCl
                                    "dichlor"

Step 4:   MeP(O)Cl2 + 2HF --> MeP(O)F2 + H2
                        (or 2NaF)   "difluor"

Step 5:   MeP(O)F2 + MeP(O)Cl2 + iPrONH2--> MeP(O)(OiPr)F
                        "Di-Di Mixture"   sarin/GB

In Step 2, the pyroester product, [MeP(O)(OH)]$_2$O, is easily formed at 300 ºC.
In Step 3, SOCl2 or PCl5 can be substituted for the chlorinating agent mixture (2PCl3 + 2Cl2).

Steps 4 and 5 can be accomplished together in the alternate reaction:

MeP(O)F2 + 2HF + iPrOH --> GB

The reactions are carried out in Pyrex glass-lined reaction vessels, except for the reaction with HF (hydrofluoric acid) or NaF (sodium fluoride), which must be carried out in Hastelloy alloy metal, or another alloy which resists hydrofluoric acid corrosion.

SYNTHESIS

Step 1:

Dimethyl hydrogen phosphite (DMHP) is prepared by auxiliary esterification, by slowly dropping in phosphorus trichloride dissolved in methanol into an ammonia gas in toluene solution, with stirring, keeping the temperature between 0-20 ºC. The methanol should be completely anhydrous, which is performed, for example, with anhydrous calcium oxide ("quick lime" or CaO; 200 g/l), by refluxing the two together for 3 hours, and then distilling off and collecting the now anhydrous methanol.

The gaseous hydrogen chloride and methyl chloride that is evolved in this reaction is removed by warming *in vacuo*, and neutralized with dry ammonia gas, and then the DMHP product is obtained purified by fractional distillation at reduced pressure (60 ºC at 3-4 mm Hg pressure).

The standard U.S. process used methyl chloride (MeCl) as the reaction mixture solvent, instead of toluene. The hydrogen chloride must be quickly removed from the reaction mixture to avoid degradation of the DMHP ester product.

Step 2:

By the thermal rearrangement/pyrolysis of DMHP, the "pyro" ester is easily formed by heating for 1 hour at 260-300 ºC, preferably in an atmosphere of nitrogen. The yield is 50-60%.

Step 3:

The dimethyl methylphosphonate ("pyro") is converted to methyl phosphonic dichloride ("dichlor") by the addition of phosphorus trichloride and dry chlorine to the cold pyromix and heated 2-3 hours at 70 ºC, followed by careful fractional distillation at 90 ºC to separate the dichlor from the phosphorus oxychloride (which boils off first), followed by careful vacuum distillation at 110 ºC/2-3 mm to separate/boil off the dichlor.

Probably uses CCl4 (carbon tetrachloride/"carbon tet") as the solvent. Methylene chloride is a good substitute for carbon tetrachloride (which was subsequently banned as a carcinogen).

Alternatively, in a flask with reflux condenser, treat 240 g of the pyromix with 400 ml thionyl chloride (SOCl2). There is no reaction in the cold mixture. Careful and light heating starts a vigorous reaction, and the pyromix dissolves slowly with the bubbling in of the SOCl2. The resulting homogenous liquid is then raised to its boiling point, and refluxed gently. Heating is slowly increased and continued until the gaseous sulfur dioxide (SO2) evolution is just complete, after 4-5 hours of reflux.

When the reaction is finished, distill off *in vacuo* (~160 mm Hg vacuum) the excess SOCl2. Continue distillation at 75 ºC/55 mm to remove the SOCl2 completely. Distil the dichlor off, at 76-76.5 ºC/43-37 mm to get 220 g of crystalline solid dichlor. Redistil at 49.5-50 ºC/11 mm to get 204 g of pure dichlor.

Step 4 & 5:

The addition of HF (hydrogen fluoride gas) or NaF (sodium fluoride), and isopropanol (isopropyl alcohol) to cold dichlor produces methylphosphonic difluoride ("difluor"), then sarin (GB), which is separated by vacuum distillation at 60 ºC/3-4 mm

Bocquet, J.R. "Ann. Soc. Roy. Sci. Med. et Nat. Bruxelles" 9:161-263 (1956). [Chem. Abs. 51: 7296 (1956)].

Campion, D.E. "The Infrared Absorption Spectra of Isopropyl Methylphosphonofluoridate and Intermediates", CW Division Scientific Report No. 7, Dugway Proving Ground (1953).

Monard, Charles and Quinchon, Jean. "Memoires Présentes á la Societé Chimique", p. 1084-1086 (1961).

Schrader, G. "The Development of New Insecticides and Chemical Warfare Agents", BIOS (British Intelligence Objectives Sub-Committee) Final Report No. 714, Item No. 8 (1947).

The DiPMP Process (Variant #1 of DMHP Process)

Step 1: $\quad$ PCl3 + 3MeOH --> (MeO)2POH + MeCl + 2HCl

Step 2: $\quad$ (MeO)2POH + MeCl + NaOMe --> (MeO)2PMe + HCl + NaOH

Step 3: $\quad$ (MeO)2PMe + SOCl2 --> MeP(O)Cl2 + SO2

You can substitute PCl5, (PCl3 + Cl2), or COCl2 for SOCl2 in this reaction.

Step 4: $\quad$ MeP(O)Cl2 + 2HF --> MeP(O)F2 + H2
$\qquad\qquad\qquad\qquad\qquad\qquad$ difluor

Step 5: $\quad$ MeP(O)F2 + iPrOH --> MeP(O)(OiPr)F
$\qquad\qquad\qquad\qquad\qquad\qquad\quad$ sarin

Synthesis

Step 1: See preceding "DMHP Process" Chapter, Step 1 of Synthesis.

Schrader, G. "The Development of New Insecticides and Chemical Warfare Agents", BIOS (British Intelligence Objectives Sub-Committee) Final Report No. 714, Item No. 8 (1947).

The IPMP Process (Isopropyl Methylphosphonite Variant #2 of DMHP Process)

Step 1:   (iPrO)P(O)(OH)Me --> [Me2P(OH)]2O + MeOMe

Step 2:   [Me2P(OH)]2O + SOCl2 --> MeP(O)Cl2 + SO2

You can substitute PCl5 or (PCl3 + Cl2) for SOCl2.

Step 3:   MeP(O)Cl2 + 2HF --> MeP(O)F2 + H2
                                           difluor

Step 4:   MeP(O)F2 + iPrOH --> MeP(O)(OiPr)F
                                           sarin

Moedritzer, "Syn. React. Inorg. Metal-Org. Chem." 4(5): 417-427 (1974).

Schrader, G. "BIOS (British Intelligence Objectives Sub-Committee) Final Report No. 714", Item No. 8 (1947).

U.S.P. #3,179,695 (1965); "Chem. Abs." 51: 7296 (1956).

U.S.P. #2,929,843 (1960).

"Mem. Poudres" 44: 119-132 (1962).

"J. Chem. Soc.", p. 1553 (1960).

"Can. J. Chem." 38: 1416-1427 (1960).

DTIC (U.S. Defense Technical Information Center) # AD0B009 338.

The ASP Process

The ASP (Aluminum Sequichloride Process) Process was never used by the U.S. because of the hazards associated with Step 1, which produced the ASP. ASP is pyrophoric, and reacts explosively with water, making it extremely dangerous to prepare and use.

Step 1:   2Al + 3MeCl --> MeAlCl2 + Me2AlCl

The 2 products together are called "Al sequichloride", or methylaluminum chloride (Me3Al2Cl3).

Step 2:   MeAlCl2 + Me2AlCl + AlCl3 --> 3MeAlCl2

Step 3:   MeAlCl2 + PCl3 --> MePCl2 – AlCl3

The product is called methylphosphonous dichloride – aluminum chloride complex.

Step 4:   MePCl2 – AlCl3 + NaCl --> MePCl2 + NaCl – AlCl3

Step 4 is the same as MAPC Process, Step 3.

Step 5:   3Cl2PMe + 1.5O2 --> 2MeP(O)Cl2

The reaction is performed in PCl3 as the solvent. The dichlor produced is converted to GB as described elsewhere.

Step 1: vapor phase catalyzed with copper metal powder; then purified by vacuum distillation at 97-101 °C/100 mm Hg.

Step 2 (alternate):   MeAlCl2 + Me2AlCl + Cl2 --> 2MeAlCl2 (86% yield) + MeCl

The reaction is performed at 0-5 °C in toluene as the solvent

SYNTHESIS

Step 1:

Finely powdered aluminum is mixed with methyl chloride in a reaction vessel with an efficient stirrer and a reflux condenser. After stirring for 1-2 hours, with a small amount of iodine to start the reaction, the reaction starts with self-heating. Stirring is continued up to 24 hours to complete the reaction. Oxygen and moisture must be rigorously excluded during the reaction.

Step 2:

Methylaluminum dichloride is prepared by adding 120 g anhydrous aluminum chloride to 177 g of the sequichloride, and the mixture heated until all the aluminum chloride was in solution. The resulting product was distilled under reduced pressure (97-100 ºC/100 mm) to give 267 g methylaluminum dichloride. This material (2.36 moles) was dissolved in 2.5 liters of gasoline, then added gradually with stirring to a solution of 325 g (2.36 moles) phosphorus trichloride in 0.5 liters of gasoline.

As the addition proceeded, the MePCl2:AlCl3 began separating out as a heavy, brown oil. After completion of the addition, the mixture is stirred overnight, then allowed to settle, and the brown oil (the bottom layer) separated off (by draining).

Step 4:

The sequichloride is heated with finely ground (to 20-40 Mesh) sodium chloride with stirring. PCl3 starts distilling first, followed by methyl phosphonous dichloride (SW) starting at 125 ºC until 165-170 ºC.

Step 5:

After charging the condenser with dry ice-acetone, pure dry oxygen (O2) gas was bubbled slowly through the methylphosphonous dichloride. An exothermic reaction started, and the temperature rose to 70 ºC. After 15 minutes, when the temperature starting to fall, external heating was applied to keep the temperature at 70 ºC.

Within 4 hours, the oxidation is complete, and the stream of oxygen stopped. The dichlor was vacuum distilled at 0.01 mm Hg into a receiving flask cooled by dry ice/acetone.

Coates, Harold and Morris, Derek, GB #1,344,051 (1974), filed 1952; U.S.P. #3,840,576 (1974), filed 1952.

Kent, Alan and Topley, Brian, GB #1,344,052 (1974), filed 1952; U.S.P. #3,829,479 (1974), filed 1955.

Zeigler, Karl and Nagel, Konrad, U.S.P. #2,744,127 (1956), filed 1953.

The U.S. Salt Process

Step 1 (esterification): PCl3 + 3iPrOH --> (iPrO)2P(OH) + iPrCl + 2HCl

The reaction to synthesize diisopropyl phosphite.

Step 2 (sodium salt): (iPrO)2P(OH) + Na(dispersed) --> (iPrO)2P(NaO) + 1/2H2

The reaction to synthesize the sodium salt of DIP

Step 3 (alkylation): (iPrO)2(NaO)P + MeCl --> (iPrO)2P(O)Me + NaCl

The reaction to synthesize the diester, diisopropyl methylphonsphonate. The reaction is carried out in a solvent of either hydrocarbon oil or ether.

Step 4 (chlorination): (iPrO)2P(O)Me + COCl2 --> Cl(iPrO)P(O)Me + iPrCl + CO2

The reaction to synthesize chloro-GB.

Step 5 (fluorination): Cl(iPrO)P(O)Me + HF --> F(iPrO)P(O)Me + HCl

SYNTHESIS

Step 1:

This is analogous to Step 1 of DMHP Process, using isopropyl alcohol instead of methyl alcohol.

Di-isopropyl hydrogen phosphite is prepared by auxillary esterification, by slowly dropping phosphorus trichloride into isopropanol dissolved in toluene, with stirring, keeping the temperature between 0-20 ºC.

The hydrogen chloride and methyl chloride evolved are removed by warming *in vacuo*, and neutralized with dry ammonia gas, and the remaining (undistilled) product purified by fractional distillation at reduced pressure (60 °C/3-4 mm). The standard U.S. Process used methyl chloride as the reaction mixture solvent. The hydrogen chloride must be quickly removed from the reaction mixture to avoid degradation of ester product.

Step 2:

265 parts of a 15% by weight dispersion of metallic sodium in a straight run petroleum distillate (b.p. 500-600 °F), and maintained at about 110 °C, and is added to a solution composed of 256 parts of di-isopropyl hydrogen phosphite in 970 parts of the petroleum distillate which has been pre-heated to about 65 °C., the rate of addition being maintained so that the temperature is at 90 °C.

Step 3:

The reaction mixture is then cooled to 85 °C and 127 parts of gaseous methyl chloride is bubbled in over a period of 25 minutes, while maintaining the temperature at between 75-90 °C. Sodium chloride in a very finely divided form precipitates as the reaction proceeds. The reaction mixture is then allowed to cool to room temperature, and the salt filtered off.

A portion of the filtrate is then mixed with about 3% by weight of diethylene glycol, and is then introduced into a vacuum distillation column operated at a pressure of 5 mm. When so operated, the desired product is taken off at about 70 °C/5 mm.

Step 4:

A stream of carbonyl chloride (phosgene/$COCl_2$) was passed into 270 g diisopropyl methylphonsphonate for 10 hours with stirring at 20-30 °C. The mixture was left overnight, and the excess phosgene and isopropyl chloride were removed at ~30 °C. Distillation *in vacuo* (38-40 °C/2 mm) gave the phosphonochloridate (chloro-GB).

Substitutes for phosgene include thionyl chloride ($SOCl_2$), PCl_5, $POCl_3$, and oxalyl chloride.

Step 5:

The crude chloro-GB (from 50 g of the diester produced by Step 3) was refluxed in 50 ml methylene chloride with 22 g sodium fluoride for 4 hours with stirring. After cooling, the whole mixture was filtered, and the residue washed twice with 20 ml of methylene chloride, which was added to the filtrate (and the residue was discarded). The methylene chloride solvent was removed by distillation under reduced pressure, and then the GB recovered by vacuum distillation at 46 °C/8 mm.

U.S.P. #3,179,690 (1965), filed in 1954, Olin Mathieson. ("Chem. Abs." 63:632g (1985))

U.S.P. #2,853,507 (1958), filed in 1953, Olin Mathieson.

U.S.P. #2,880,224 (1959), filed in 1953, Olin Mathieson.

"J. Chem. Soc.", p. 1553-1555 (1960), Bryant, P.J.R. and Ford-Moore, A.H.

"J. Chem. Soc.", p. 238-240 (1961).

"Helv. Chim. Acta" 56:492-494 (1973), Maier, Ludwig.

The TIPP Process

The TIPP (Tri-isopropyl phosphite) Process:

Step 1:   PCl3 + 3iPrOH + 3NH3 --> (iPrO)3P + 3NH4Cl

The reaction is carried out in a hydrocarbon solvent. The ammonia gas performs the function of a tertiary base to enable this esterification reaction.

Step 2:   (iPrO)3P + MeI + (iPrO)2PMe + iPrOI

The reaction synthesizes diisopropyl methylphosphonite.

Step 3 }
        } ---- same as Steps 4 and 5 of U.S. Salt Process
Step 4 }

SYNTHESIS

Step 1:

You must use absolute (anhydrous) isopropanol, such as prepared by distilling over sodium metal. To a solution of 180 g (3 moles) of anhydrous isopropanol, and 447 g (477 ml/3 moles) of freshly distilled N,N-diethylaniline (Danger! Extremely poisonous!) in 1 liter of dry petroleum ether (b.p. 40-60 ºC) is added drop-wise a solution of 137.5 g (87.5 ml/1 mole) phosphorus trichloride in 400 ml dry petroleum ether (b.p. 40-60 ºC).   The flask is kept cooled in a cold water bath.

With vigorous stirring, the PCl3 is added at a sufficient rate that the mixture boils gently towards the end of the addition (about 30 minutes). After the addition is complete, the mixture is heated under gentle reflux for about 1 hour with stirring. It is then cooled, and then suction filtered through a sintered glass funnel, to remove the copious amount of diethylaniline hydrochloride that has precipitated out.

The filtered salt is compressed and the liquid saved, and the salt cake is washed with five 100 ml portions of petroleum ether (b.p. 40-60 ºC). The filtrate and five washings are combined, and then purified by first distilling at water-bath temperature through a 75 cm tall Vigreux fractional distillation column. Then the residue is then distilled under water pump vacuum, through a 75 cm Vigreux column. After distilling over a small fore-run (which is discarded), the desired product is collected at 43.5 ºC/1 mm.

The synthesis has been done with anhydrous ammonia gas (NH3), pyridine, triamylamine, or ammonium carbamate, in place of the (extremely poisonous) tertiary amine, N,N-diethylaniline.

Step 2:

In a 2-liter, round-bottom flask is added 284 g (113 ml/2 moles) of methyl iodide. A water-cooled reflux condenser and a dropping funnel is then fitted on. The dropping funnel is charged with 416 g (435 ml/2 moles) triisopropyl phosphite. A few pieces of porous ceramic plate are added to the methyl iodide, and about 50 ml of the phosphite is added. The mixture is heated gently, until an exothermic reaction begins. The heating is then stopped, and the remainder of the phosphite is added at such a rate that the mixture keeps boiling briskly. Towards the end of the addition, it may be necessary to resume heating. After the addition is complete, the mixture is now boiled under reflux for 1 hour.

The condenser is replaced by a 50-75 cm Vigreux column, attached to a condenser set up for distillation, and most of the isopropyl iodide is distilled off at 85-95 ºC (at atmospheric pressure). The residue is distilled *in vacuo* (water-pump pressure) with a dry ice trap between the distillation receiver and the pump. 310 g (91%) of isopropyl iodide is recovered. The rest is distilled *in vacuo* at vacuum pump pressure. Except for a small fore-run (which is discarded), the colorless product distills at 51 ºC/1 mm. The yield is 308-325 g (85-90%).

Rabjohn, N., ed. "Organic Synthesis, Collective Volume 4". NY: John Wiley (1963).

U.S.P. #2,678,940 (1954), filed 1951.

U.S.P. #2,848,474 (1958), filed 1953; Monsanto.

U.S.P. #2,859,238 (1958), Monsanto.

The TIPP Process (Alternate #2)

The TIPP (Tri-isopropyl phosphite) Process:

Step 1:   PCl3 + 3iPrOH + 3NH3 --> (iPrO)3P + 3NH4Cl

The reaction (which synthesizes TIPP) is performed in hydrocarbon solvent. The ammonia gas (NH3) is the catalytic tertiary base for the reaction.

Step 2:   (iPrO)3P + NaI + (iPrO)2PMe + iPrOI

The reaction to synthesize diisopropyl methylphosphonite.

Step 3:   (iPrO)2PMe + PCl5 --> MeP(O)Cl2

The reaction to synthesize dichlor.

Step 4:   MeP(O)Cl2 + 2HF --> MeP(O)F2 + H2

The reaction to convert dichlor to difluor.

Step 5:   MeP(O)F2 + iPrOH --> MeP(O)(OiPr)F

The final step to synthesize sarin (GB).

SYNTHESIS

Step 1: [See preceding "TIPP Process" Chapter]

Step 2:

83.5% yield was obtained from TIPP, and a catalytic amount of sodium iodide, to get MeP(OiPr)2.

Step 3:

A mixture of 116 g MeP(O)(OiPr)2 in 120 ml POCl3 and 131 g PCl3 was treated with chlorine gas for 3 hrs. Fractional distillation gave 84.2 g (75.4 mole-%) of dichlor.

Steps 4 & 5: See elsewhere.

The TMP Process

The TMP (trimethyl phosphite) Process:

Step 1:   PCl3 + 3 MeOH --> (MeO)3P   )
)--- analogous to TIPP
Step 2:   (MeO)3P + MeI --> (MeO)2PMe + MeOI   )   Process

Step 3:   (MeO)2PMe + COCl2 --> (MeO)P(Me)(Cl)   )
)--- analogous to Steps 4 & 5 of
Step 4:   (MeO)P(Me)(Cl) + NaF + iPrOH --> GB   )   U.S. Salt Process

SYNTHESIS

Step 1: [See preceding "TIPP Process" Chapter]

Step 2: [See preceding "TIPP Process" Chapter]

Step 3: [See "U.S. Salt Process" Chapter]

Step 4: [See "U.S. Salt Process" Chapter]

The HTM-Catalytic Process

The HTM-Catalytic (High Temperature Methanation - Catalytic) Process:

Step 1:   PCl3 + CH4 --> MePCl2 + HCl

The reaction is carried out at 550-650 °C.

Step 2:   2MePCl2 + O2 --> MeP(O)Cl2

Step 3:   MeP(O)Cl2 + 2HF --> MeP(O)F2 + H2

Step 4:   MeP(O)F2 + iPrOH --> MeP(O)(OiPr)F

The advantage of this process over the DMHP Process is the lack of the latter's POCl3 by-product, which is difficult to deal with/purify out.

SYNTHESIS

Step 1:

The reactor consisted of a quartz tube of 24 mm outside diameter, 40 cm long, heated by an annular furnace 33 cm long. The heated section of the tube is packed with inert packing material such as broken quartz (4 mesh screen size). Using the reactor, methane gas (CH4) was passed through a reservoir of phosphorus trichloride at a temperature to give a gaseous mixture of 1.88 moles PCl3 and 0.11 moles of catalytic O2 gas, which was then passed through the reactor heated to a temperature of 575 °C in the reaction zone, at a residence time of 0.2 seconds.

The methyl phosphonous dichloride (SW) was produced in a 24% yield and is separated by distillation at 115 °C.

Step 2:

See "MAPC Process," step 3

Step 3 & 4:

See elsewhere for converting dichlor to sarin.

Pianfetti, John A. and Quin, Louis D. "JACS" 84: 851-854 (1962).

U.S.P. #3,210,418 (1965), filed 1953; Pianfetti, FMC Corporation.

The HTM-Pyro Process

The High Temperature Methanation-Pyro Process:

Step 1: $PCl_3 + 3MeOH \rightarrow (MeO)_2POH + MeCl + 2HCl$

                     heat
Step 2: $(MeO)_2POH \rightarrow [MeP(O)(OH)]_2O + Me_2O + (MeO)(OH)P(O)Me$

The $[MeP(O)(OH)_2)$ is known as "pyro". The mixture of the three products of the reaction is known as "pyromix". Step 1 & 2 are the same as the DMHP Process (see earlier).

Step 3: $PCl_3 + CH_4 \rightarrow MePCl_2 + HCl$

The reaction is carried out at 550-650 °C.

Step 4: $[MeP(O)OH)]_2O + 3MePCl_2 + 3Cl_2 \rightarrow 3MeP(O)Cl_2 + 2HCl$

The step uses together the products of Step 2 & 3 to produce dichlor.

Step 5: $MeP(O)Cl_2 + 2HF \rightarrow MeP(O)F_2 + H_2$

Step 6: $MeP(O)F_2 + iPrOH \rightarrow MeP(O)(OiPr)F$

Synthesis

See elsewhere for the synthetic details of Steps 1 – 6.

The APC Processes

The APC (Aluminum-Phosphorus-Chloride) Process:

Step 1:   PCl3 + anhyd. AlCl3 + MeCl --> MeCl:PCl3:AlCl3

The reaction is performed in methylene chloride solvent.

Step 2:   MePCl:AlCl4 + DEP --> MePCl4 + DEP.AlCl3

                         MePCl4 + SO2 --> CH3P(O)Cl2 + SOCl2

Step 3:   MeP(O)Cl2 + 2HF --> MeP(O)F2 + H2

Step 4:   MeP(O)F2 + iPrOH --> MeP(O)(OiPr)F

The APC Process produces a high purity product. It's attractive to produce a run of a few hundred tons. It's easy to produce this "limited amount" of dichor (DC). However the requirement for the recovery of aluminum chloride hexahydrate made the process uncompetitive.

SYNTHESIS

Step 1:

Good yields are dependent on the complete formation of the APC (aluminum chloride – phosphorus trichloride – methyl chloride) complex.

To prepare the APC complex, 200 g (1.5 moles) of very finely powdered, anhydrous aluminum chloride, 137.5 g (1 mole) of phosphorus trichloride, and 50.5 g (1 mole) of methyl chloride are used. The PCl3 is run into the AlCl3 powder, and the very thick mixture is heated to 70-75 ºC, and then agitated with a "vibrator" type stirrer (such as a "Vibromix" stirrer).

A cold finger-type, acetone/dry ice condenser is used as the input source of methyl chloride, which is slow at first. After about an hour, the methyl chloride absorption increases markedly, and stays that way for 2.5 to 3 hours, the reaction mixture becoming very fluid. At the end of this stage the reaction mixture becomes more viscous, and absorption slows with addition being reduced. After 2 hours methyl chloride addition is stopped, but heating and stirring are allowed to continue for another hour. They are then both stopped, and the cooling condenser turned off.

Throughout the entire reaction, the addition of methyl chloride addition must be carefully controlled, with the reaction temperature being maintained at 70-75 ºC.

Step 2 (Dichlor from APC Complex):

Treatment with DEP (Diethyl Phthalate), followed by SO2 gives 80-85% dichlor on vacuum distillation.

The reaction flask in which the APC Complex is prepared is also used for the breakdown of the complex and as the distillation flask for the recovery of the crude dichlor, 3 moles per mole of complex of dry diethyl phthalate is added to the complex. Stirring is begun to dissolve the complex as much as possible in the DEP. Dissolution is accompanied by a pronounced rise in temperature, and external water cooling should be used in the early stages, until the solid breaks up. At this point a stream (10-12 liters/hour) of dry sulfur dioxide (SO2) gas is passed through the stirred mixture. An exothermic reaction occurs and the temperature of the reaction mixture normally rises to a maximum of 50 ºC, and after 2 hours begins to fall. Sulfur dioxide is passed into the stirred mixture for a further 30 minutes, and the mixture is then allowed to stand for a further 1 hour to ensure completion of the reaction.

The receiver is immersed in dry ice/acetone cooling mixture, and the apparatus evacuated to a vacuum of approximately 0.1 – 0.5 mm Hg, by a high capacity high-vacuum pump. The distillation flask is heated to 85 - 90 ºC., whereupon the dichlor and residual thionyl chloride are separated/distilled off from the DEP. After 6 hours the distillation is considered complete.

The receiving flask now becomes the distillation flask to separate the thionyl chloride (SOCl2) and dichlor. Heating is continued until the temperature is approximately 150-155 ºC, which indicates that the SOCl2 has distilled off, and the dichlor is about to distil. Heating is stopped and the apparatus allowed to cool to room temperature.

The apparatus is modified by addition of a water vacuum pump and a fine capillary bleed. The apparatus is evacuated to 60 mm Hg using a water pump. During this procedure the last traces of SOCl2 are removed. At 60 mm Hg, the dichlor distils at 84-85 ºC.

Step2 3 & 4:

See elsewhere.

Reesor, J.B. et al. "Can. J. Chem." 38:1416-1427 (1960).

Hignett, T.P. et al. U.S.P. #2,875,245 (1959), filed 1955. ["Chem. Abs." 53:12177A (1959)]

Clay, John P., "J.Org.Chem" 16: 892-894 (1951).

Jauhiainen, T.P. and Lindberg, Johann. "Finn. Chem. Letters", p. 18-22 (1977).

The MAPC Process

The Modified APC Process:

Step 1:   PCl3 + anhydrous 2AlCl3 + MeCl --> Cl2PMe:AlCl4

Step 2:   CH3PMe:AlCl4 + NaCl --> MePCl2 + NaCl

Step 3:   2MePCl2 + O2 --> 2 MeP(O)Cl2

SYNTHESIS

Step 1:

Same as step 1 of the APC Method

Step 2:

The complex was heated with finely ground (to 20-40 Mesh) sodium chloride with stirring. PCl3 starts distilling first, then the methylphosphonous dichloride (SW) starts at 125 °C until 165-170 °C.

Step 3:

The freshly prepared SW was run into a flask from a dropping funnel. With a dry ice/acetone cold finger condenser attached, pure, dry oxygen gas was slowly bubbled into the SW. An exothermic reaction starts, and the temperature rises to 70 °C. After 15 minutes, it requires external heating (for 4 hours) to maintain the reaction flask at 70 °C. Then the mixture was distilled in vacuo (0.01 mm Hg) to give 95% yield of dichlor. [See elsewhere for synthesis of GB from dichlor.]

de Borst, C. et al. "The preparation of radioactively labelled organophosphorus compounds on a semi-microscale", Report 1975-13, Chemisch Laboratorium TNO, the Netherlands (1974).

Hechenbleikner et al., U.S.P. #4,411,852, (1983), "Oxidation of Alkyl Phosphonous Dichlorides".

Kent & Topley, U.S.P. #3,829,479 (1974), filed 1955. "Preparation of alkyldichlorophosphines".

Perry, B.J. et al. "Can. J. Chem." 41: 2299-2302 (1963) [B.P. #1,346,410 (1974)].

Soroka, Miroslaw. "A Simple Preparation of Methyl Phosphonous Dichloride", "Synthesis", p. 450 (1977).

Dichlor from Dimethyl Methylphosphonate

Step 0:   Step 1 of HTM-Catalytic Process to prepare SW (MePCl2)

Step 1:   MePCl2 + 2 MeOH → MeP(OMe)2

Step 2:   MeP(OMe)2 + SOCl2 → MeP(O)Cl2

SYNTHESIS

Step 1:

55 g (0.5 moles) methyl phosphonous dichloride in 50 ml pentane was added slowly over a period of 30 minutes to a stirred solution of 35.2 g (1.1 moles) of methanol in 350 ml pentane. The temperature was maintained in the range of 20-25 ºC. The pH of the reaction mixture, as indicated by a pH meter with a dripping KCl electrode, was held in the range of 7.0-8.5 by the addition of anhydrous ammonia gas. Upon completion of the addition, stirring was continued for 30 minutes at 0-5 ºC. The reaction mixture was then washed with 800 ml of 3.5% NaOH solution and then dried over anhydrous sodium (or magnesium) sulfate. The product was freed of solvent by distillation, and then the product was distilled to give O,O-dimethyl methylphosphonite.

Step 2:

A mixture of 124 g (1 mole) of O,O-dimethyl methylphosphonite and 0.73 g of N,N, dimethylformamide was added drop-wise at reflux temperature, in the course of 2 hours to 297.5 g (2.5 moles) of thionyl chloride. An intense generation of sulfur dioxide and methyl chloride occurs during the drop-wise addition. After completion of the addition, stirring is maintained for 7.5 hrs. at reflux temperature. Unreacted thionyl chloride is subsequently evaporated off at 25 ºC under a water-jet vacuum.

There is obtained 133.1 g of crude dichlor, which crystallizes completely on standing at room temperature. Vacuum distillation of the crude dichlor at 56-57 ºC/14 mm Hg yields 125.5 g (94% yield) of pure dichlor having a m.p. of 33 ºC. [See elsewhere for synthesis of GB from dichlor.]

Maier, Ludwig. U.S.P. #4,213,922 (1980).

Harowitz, Charles. U.S.P. #2,903,475 (1959).

Dawson, Thomas et al. U.S.P. #2,847,469 (1958), Army.

The Methylphosphonic Acid Method

Step 1:   MeCl + PCl3 + AlCl3 --> MePCl4:AlCl3

Step 2:   MePCl4:AlCl3 + 9H2O --> MeP(O)(OH)2 + 2AlCl3:6H2O + 4HCl

Step 3:   MeP(O)(OH)2 + NaF --> MeP(O)F2

KF or HF can also be used instead of NaF. The reaction is carried out in dilute hydrochloric acid. Dilute HNO3 can be substituted for the dilute hydrochloric acid.

Step 4:   MeP(O)F2 + iPrOH + isopropylamine --> sarin

SYNTHESIS

Step 1:

15 ml methyl chloride, 22 ml phosphorus trichloride, and 40 g powdered anhydrous aluminum chloride are introduced into a thick-walled flask, securely stoppered (ground glass stopper!), and the flask shaken for a few hours, after which the AlCl3 has disappeared and the complex, originally liquid, has solidified.

Step 2:

The solid is dissolved in methylene chloride, and the solution treated with 60 ml of water, which is added very slowly, the temperature not being allowed to exceed -5 ºC. The reaction mixture is filtered, and the solvent evaporated from the filtrate.

Perren, Edward Arthur, and Kinnear, Alan Macpherson, B.P. #707,961 (1954), filed 1948.

The GB Alternate #1 Method

Step 1:   2PCl3 + 2HF ---> 2FPCl2 + Cl2

Step 2:   (F)PCl2

GB Precursors

Red Phosphorus

White phosphorus is produced by thermal reduction of monocalcium phosphate in a coventional rotary kiln, by admixing the phosphate with a molar excess of charcoal, adding silica (common sand, SiO_2), forming the mixture into pellets, and heating them rapidly to a temperature of about 2,100 to 2,400 °F for not longer than 15 minutes.

You can also use triple superphosphate or even clean animal bones.

SYNTHESIS

The reaction apparatus to be used requires a trip to the plumbing or hardware store. The "furnace" is a plumbing nipple (a short length of tubing about 3" in diameter and 6" long threaded on both ends. An end cap is screwed on one end of the furnace nipple. At the other end is screwed a 1" reducer, and a 90° elbow after that. Finally, a length of 1" tube is attached. It will be submerged in a beaker full of water once the reaction starts, as white phosphorus is produced which is inflammable on exposure to air. The furnace is heated by a couple of propane torches on full.

100 g of fertilizer grade triple superphosphate was intimately mixed with 24 g of ground charcoal, 13 g sand, and 40 ml of water. The mixture was pressed into pellets and pressed with a 10 ton hydraulic press. The green pellets were dried overnight at about 220 °F.

A handful of pellets was placed in a furnace and heated to 2,225 °F for 2 hrs.

When you turn off the furnace heat, immediately remove the pipe from under the surface of the water-bath to avoid the water suck-back from its cooling, and the resulting dangerous high pressure steam blow-out.

The white phosphorus produced is catalytically converted to red phosphorus by heating with a speck of iodine crystal.

Dancy, U.S.P. #3,923,961 (1975)

Phosphorus Trichloride (PCl3)

Phosphorus trichloride is a poisonous, colorless, fuming liquid (b.p. 73.5 ºC; and density of 1.57 g/cm3) with a pungent smell. It's prepared by the action of a stream of dry chlorine gas on white phosphorus on heating in a ground-glass jointed distillation setup or a simple retort, and condensing the PCl3 vapor produced in a cooled, dry receiver. Do not use an excess of chlorine gas, to avoid producing PCl5:

$$P_4 + 6Cl_2 \rightarrow 4PCl_3$$

2.4 kg white phosphorus and 8.15 kg chlorine gas produce 10 kg PCl3.

White phosphorus can be prepared from red phosphorus by subliming the red phosphorus by gentle heating it, and contained in a large flask. Or white phosphorus can be produced by the method previously described earlier.

Alternate method:

$$Ca_3P_2 + Cl_2 \rightarrow PCl_3 + CaCl_2$$

The calcium phosphide (Ca3P2) is prepared easily by roasting calcium phosphate with charcoal at around 1,000 ºC. Use a steel pipe with threaded ends and sealed with 2 screw-on endcaps.

PCl3 reacts violently with water, releasing corrosive and poisonous hydrogen chloride gas. It is soluble in (diethyl) ether and chloroform (1 g P4 in 40 ml chloroform).

Phosphorus Oxychloride (POCl3)

POCl3 can be made by various methods:

1. The hydrolysis of phosphorus pentachloride (PCl5):

$$PCl_5 + H_2O \rightarrow POCl_3 + 2\ HCl$$

2. The reaction of phosphorus trichloride with pure oxygen gas (air is ineffective) at 20–50°C:

$$2\ PCl_3 + O_2 \rightarrow 2\ POCl_3$$

3. The reaction of phosphorus pentachloride with boric acid or oxalic acid:

$$3\ PCl_5 + 2\ B(OH)_3 \rightarrow 3\ POCl_3 + B_2O_3 + 6HCl$$
$$PCl_5 + (COOH)_2 \rightarrow POCl_3 + CO + CO_2 + 2\ HCl$$

4. The oxidation of phosphorus trichloride with potassium chlorate:

$$3\ PCl_3 + KClO_3 \rightarrow 3\ POCl_3 + KCl$$

5. The reaction of phosphorus pentoxide with sodium chloride:

$$2\ P_2O_5 + 3\ NaCl \rightarrow 3\ NaPO_3 + POCl_3$$

6. The heat reduction of tricalcium phosphate with carbon in the presence of chlorine gas:

$$Ca_3(PO_4)_2 + 6\ C + 6\ Cl_2 \rightarrow 3\ CaCl_2 + 6\ CO + 2\ POCl_3$$

V-Agents

VX is the most toxic of all V-agents, and the agent stockpiled by the U.S. by the late 1950's to the end of the 1980's. It is much more toxic than G-agents, as well as what is called a "persistent" nerve agent, a thick liquid, low volatility, long-lasting agent when dispensed, unlike unthickened G-agents which disperse much more readily.

It was more or less simultaneously and independently discovered by three chemists in three different countries in the early 1950's: Dr. R. Ghosh of ICI (Imperial Chemical Industries) in England, Dr. Gerhard Schrader in Germany, and Dr. Lars Eric Tammelin in Sweden. Dr. Ghosh was first, synthesizing VX in 1952.

Its formula is: $MeP(O)(OEt)(S-Et-N{<}{iPr \atop iPr}$

The Newport Process VX precursor QL not particularly stable and reacts violently with air or water.

The Newport Process

The Newport Process was named after the Newport, Indiana production plant where V-agents were first made in a large scale in the U.S. in the mid-1950's:

Step 0:   Step 1 of HTM-Catalytic Process to prepare SW (MePCl2)

Step 1:   MePCl2 + 2EtOH --> MeP(OEt)2

Step 2:   MeP(OEt)2 + (iPr)2NEtOH --> QL

QL = Ethyl 2-diisopropylaminoethyl methylphosphonite

Step 3:   QL + S --> VX

Industrial Process:

Step 1: Uses isobutane as the solvent (under pressure).

Step 2: Uses glacial acetic acid (AcOH) as the solvent.

SYNTHESIS

Step 1:

1 liter of diethyl ether, previously dried over sodium, and 234 g (2 moles) of dichlorophosphine (SW; MePCl2) were added to the reaction flask that had previously been flushed with dry nitrogen gas.

A mixture of 193.2 g (4.2 moles) of absolute ethanol and 627 g (4.2 moles) of N,N-diethylaniline was added drop-wise through a dropping funnel, with stirring. During addition of the alcohol, the reaction temperature was maintained between 20-30 ºC, using an ice bath, the system was periodically flushed with dry nitrogen, and the exit gas line of the condenser was connected to a mercury bubbler. After the alcohol was completely added, the reaction mixture was stirred for an additional 3 hours. The flask was removed from the apparatus and flushed with nitrogen, and then the contents poured into a Buchner funnel.

The filtrate was transferred to a nitrogen-flushed 2 liter flask, connected to a 10" packed distillation column, and the ether distilled off at 60 ºC. During the removal of the ether, the exit gas line was sealed by a mercury bubbler to prevent entry of oxygen into the system.

Distillation *in vacuo* gave 223 g of diethyl methylphosphonite (47 ºC/50 mm Hg) (80%yield).

Step 2:

75 g of diethyl methylphosphonite was mixed in a round-bottom flask, and mixed with 40.2 g (0.28 moles) of 2-diisopropylaminoethanol. The reaction flask was then flushed with nitrogen and heated slowly with a Glas-Col heating mantle for 55 minutes from 23 °C to 110 °C, the reflux temperature of the reaction mixture. The ethanol formed in the reaction was continuously removed, with the still-head temperature varying from 75-78.5 °C. An additional 65 minutes was necessary to complete the removal of the ethanol, and the temperature was finally 150 °C.

Heat was discontinued and dry nitrogen was flushed through the apparatus, while the temperature cooled to 50 °C, and then distilled *in vacuo* (54 °C/0.1 mm Hg) to collect ethyl diisopropylaminoethyl methylphosphite.

Step 3:

An ethylene glycol bath is used either for heating, or (with dry ice) cooling. The reaction flask was charged with 3,221 g (1.37 moles) QL, and a nitrogen gas purge was then started to maintain an inert atmosphere in the flask. Then 435 g (1.36 moles) of flowers of sulfur was slowly added. The heat of reaction was removed by the dry ice cooled bath, and the temperature kept at approximately 30 °C. It took approximately 1 hour to add all the sulfur. After ten minutes, to allow completion of the reaction, the flask was heated as quickly as possible to 120 °C., and maintained at this temperature for 90 minutes.

The reaction flask was allowed to cool, and 98% VX was drained.

Eckhaus, Sigmund R. U.S.P. #3,911,059 (1975), filed 1960.

Ford-Moore, Arthur Henry and Bebbington, Alan, B.P. #1,375,690 (1975), filed 1957 ("Chem. Abs." 82: P171204S (1975).

The Water Method for VX

Step 0:   MePCl2 + 2EtOH --> MeP(O)(OEt)2

Step 1:   MeP(O)(OEt)2 + S --> MeP(S)(OEt)2

Step 2:   MeP(S)(OEt)2 + H2O --> MeP(S)(OEt)(OH)

Step 3:   MeP(S)(OEt)(OH) + NaOH --> MeP(O)(OEt)(SNa)

Step 4   MeP(O)(OEt)(SNa) + (iPr)2>NEt Cl:HCl --> VX

SYNTHESIS

Step 2:

The acid ester prepared from the diethyl ester is dissolved with cooling in aqueous NaOH containing one equivalent of base. One mole diisopropylaminoethyl chloride hydrochloride and one mole aqueous NaOH are added, the mixture placed in a continuous extraction (Soxhlet) apparatus of high efficiency and allowed to stand at room temperature for a few hours for the components to react. The mixture is then extracted with toluene for several hours, and the toluene extract worked up in the usual way for the separation of the product.

Step 3:

Dissolve ethyl hydrogen methylphosphonothionate in 1 molar equivalent of aqueous sodium hydroxide, and allow the mixture to stand for a few hours. Then extract the mixture with boiling toluene, separate off the toluene extract (you can recover the toluene for reuse by distillation), and distill the residue *in vacuo* to get VX.

Epstein, Joseph. U.S.P. #3,903,210 (1975), filed 1958.

Ford-Moore & Bebbington, U.S.P. #3,781,387 (1973), filed 1956.

Volkova et al. AD406 972, JPRS (1961).

Webster, Harold, U.S.P. #3,035,082 (1962), filed 1959.

VX from Ethyl MethylPhosphonothiolate

Step 0:   Step 1 of HTM-Catalytic Process to prepare SW (MePCl2)

Step 1:   MeP(OEt)(O)H

Step 2:   MeP(OEt)(O-) + S --> [MeP(S)(OEt)(O-)]

Step 3:   [MeP(S)(OEt)(O-)] + (iPr)2>NEtCl+ --> VX + HCl

SYNTHESIS

Step 2 & 3:

A mixture of 1 mole of flowers of sulfur and 1 mole diisopropylaminoethyl chloride is vigorously stirred in 100 ml toluene. While vigorous stirring continues, the whole thing is maintained at 40 °C and 1 mole of ethyl hydrogen methylphosphonite is added at such a rate that the temperature is maintained at 35-40 °C.

Heating is continued for 1 hr., and then the whole thing is allowed to cool to room temperature.

The resulting semi-crystalline mass is then treated with water, and the product free base liberated with aqueous sodium carbonate, extracted with toluene, the toluene removed by distillation, and then the VX isolated by distillation in vacuo to give a yield of 77% of VX.

U.K. Patent #1,346,609 (1974), filed in 1962, Ley and Sainsbury, UK Government.

Koblin, U.S.P. #4,708,869 (1987), filed 1969, Dept. of the Army.

VX From Methylphosphonothioic Dichloride

Step 1:   MePCl2 + EtOH + tert. base + S --> MeP(S)(OEt)(Cl) + HCl

Step 2:   MeP(S)(OEt)(Cl) + NaOEtN<(isoPr)2 + heat --> VX

SYNTHESIS

Step 1:

A mixture of 149 g (1 mole) methylphosphonous dichloride and 400-600 ml dry toluene or gasoline was stirred, while a slow stream of dry nitrogen was passed through the liquid phase during the entire run. To the stirred solution was added with cooling at an internal temperature of 10-20 ºC. a mixture of one mole each of triethylamine or diethylaniline and dry ethanol. The resulting slurry of amine HCl was stirred for 1-2 hours, allowing the temperature of the reaction mixture to rise slowly to room temperature. The vigorously stirred reaction mixture was treated with the calculated amount of sulfur flowers in several portions at such a rate as to maintain a reaction temperature of 25-35 ºC. After completion of the addition, 150-200 ml distilled water was added and the mixture stirred until the amine hydrochloride had passed into solution. The contents of the reaction flask was suction filtered to remove the unreacted sulfur. The organic phase of the filtrate was washed with 1-1.5 N hydrochloric acid to remove any unreacted amine. After drying over anhydrous magnesium sulfate, the solvent was removed from the solution under reduced pressure. The crude liquid product was distilled in vacuo to yield the product, 837 g (53% yield) O-ethyl methylphosphonochloridothioate.

Step 2:

69 g of diisopropylaminoethanol is added with stirring at 50 ºC to a suspension of 6 g metallic sodium powder in 50 ml of toluene. The sodium is dissolved after 2 hrs. 45 g of O-ethyl methlphosphonochloridothioate is added dropwise at 50 ºC., the mixture is stirred for an hour, then diluted with 200 ml toluene, and washed with 250 ml of water. The toluene layer is separated and dried over anhydrous sodium sulfate, followed by distilling off of the toluene to leave a residue of VX.

Acetone or chloroform may be substituted for the toluene solvent.

Hoffmann, Friedrich et al. "JACS" 80:3945-3948 (1958)

Schegk, Ernst et al. U.S.P. #3,014,943 (1961)

VX from the Methylphosphonothioic Chloride and DIIPAE

Step 0:   MeP(S)(OEt)(Cl) is prepared as already described.

Step 1:   MeP(S)(OEt)(Cl) + HO-Et-N<(isoPr)2 --> VX + HCl

SYNTHESIS

Step 1:

A mixture of 18.5 parts of ethyl methylchlorothionophosphonate, 12 parts of diisopropylaminoethanol, 10.6 parts anhydrous sodium carbonate, 0.2 parts copper-bronze powder, and 300 parts of xylene are heated at 90 ºC for five hours. The mixture is filtered of NaCl and distilled in vacuo.

Ghosh, U.S.P. #2,863,901 (1958), filed in 1953

VX from the Methylphosphonothioic Dichloride and the Mercaptan

Step 0:   MeP(O)(OEt)(Cl) is prepared as already described in the U.S. Salt Process.

Step 1:   MeP(O)(OEt)(Cl) + HS-Et-N<(isoPr)2 --> VX + NaCl

SYNTHESIS

Step 1:

0.25 moles of β-diisopropylaminoethyl mercaptan dissolved in 50 ml toluene are added dropwise with stirring at 40 to 50 °C to a suspension of metallic sodium in 100 ml of toluene. The sodium is dissolved after about 30 minutes. 40 g methylphosphonic acid ethyl ester is then added with further stirring at 40 °C. The mixture is heated to 50 °C for another hour, then cooled to room temperature and about 2-3 ml of water added to the reaction product. The NaCl formed can now be filtered off. The filtrate is dried over sodium sulfate and then fractionated in vacuo.

Schegk, Ernst et al. U.S.P. #3,014,943 (1961)

Lars-Eric Tammelin, Acta Chem. Scand. 11(8):1340-1349 (1957)

VX from Dichlor via the Mercaptan

This is an unpublished [i.e. new] synthesis for VX that seems to have been overlooked by U.S. Chemical Weapons Center chemists.

Step 1:   MeP(O)(OR)2 + SOCl2 --> MeP(O)Cl2

Step 2:   MeP(O)Cl2 + Na/EtOH + H2S --> MeP(O)(OEt)SH

Step 3:   MeP(O)(OEt)SH + ClEtN<(iPr)2 --> VX + HCl

SYNTHESIS

Step 1:

7g (0.056 moles) of dimethyl methylphosphonolate was placed in a reaction vessel with magnetic stirring, and an anhydrous calcium chloride guard tube. 20 g (0.17 moles) thionyl chloride was added dropwise with stirring over a period of 30 minutes at room temperature. After the addition was complete 0.5 ml pyridine was added and the reaction mixture refluxed with stirring for 4 hours in an oil bath. The excess thionyl chloride was then removed by distillation. The product, dichlor, was then distilled at 158-161 ºC at atmospheric pressure. Yield: 6.5g (87%).

Step 2:

100 ml dry ethanol was added to a 250 ml three-necked round bottom flask. 3.1g (0.135 moles) sodium metal was added in small pieces and dissolved at room temperature. The reaction mixture was then cooled to 0-5 ºC using an ice-salt cooling bath. Dry hydrogen sulfide as then passed through the reaction mixture till it was saturated. 6 g (0.0045 moles) of dichlor was then added dropwise over a period of 3 hours at -5 ºC with stirring. The ice bath was then removed and the reaction mixture was stirred at room temperature, followed by reflux for 4 hours. Excess ethanol was distilled off, the crude product was dissolved in the minimum amount of water and extracted with diethyl ether. The ether layer was discarded and the aqueous layer was acidified by conc. hydrochloric acid up to pH ~2. It was then extracted with diethyl ether repeatedly. The ether layer was dried over anhydrous magnesium sulfate, and the solvent removed by distillation. The residue, ethyl methylphonsphonothioc acid was distilled in vacuo. Yield: 4.7g (75%), b.p. 92º/1 mm Hg

Step 3:

A sodium ethylate solution containing 0.17 moles of dissolved metallic sodium is diluted with 100 ml of toluene. At 20 ºC 24g (0.17moles) of β-diisopropylaminoethyl chloride is added dropwise with stirring. 50 g ethyl methylphonsphonothioic acid. After the reaction is complete, the reaction mixture is refluxed on a water bath. It is then diluted with toluene and the NaCl filtered off. After distilling in vacuo, you collect the VX.

Calderbank & Ghosh, U.K. Patent #832,990 (1960).

"Indian J. Chem." 50B:1504-1509 (2011)

Schegk, Ernst et al. U.S.P. #3,014,943A.

Shan, Y., unpublished.

VG from Phosphorus Pentasulphide and the Mercaptan

Step 1:   4P + 10S --> 2P2S5

The reaction is carried out at 300 ºC.

Step 2:   P2S5 + 2EtOH --> (EtO)2P(S)SH   [diethyl dithiophosphoric acid]

Step 3:   (EtO)2P(S)SH + ClEtN<(Et)2 ---> (EtO)2P(S)(OEtN<(Et)2)

Step 4 (isomerization): (EtO)2P(S)(OEtN<(Et)2) --> (EtO)2P(O)(SEtN<(Et)2)

SYNTHESIS

Step 1:

Phosphorus pentasulfide is made by the reaction of liquid white phosphorus with sulfur at 300 ºC.

Step 4:

Heat the product from Step 3 to 120 ºC., and maintain at this temperature for 90 minutes.

VX via the Thiocyanate

Step 1:   MeP(OEt)2 + NCSEtN<iPr2 --> VX

SYNTHESIS

Step 1:

33 parts of diisopropylaminoethyl thiocyanate are added cautiously to 33 parts of diethyl methylphosphonolate on a steam bath, maintaining the temperature below 107 ºC. The mixture is then left overnight and then heated at 120-125 ºC under 60-65 mm vacuum for 1 hour. The product is isolated by distillation in vacuo.

The thiocyanate may be made by interacting diisopropylaminoethyl chloride with potassium thiocyanate in acetone at room temperature.

Ghosh, U.K. Patent #763,516, ICI (1956)

Alternate VX Process #7

Step 0:   Step 1 of HTM-Catalytic Process to prepare SW (MePCl2)

Step 1:   MePCl2 + EtOH + S + NH3 --> MeP(OEt)(O)(S NH4-)

Step 2:   MeP(OEt)(O)(S NH4-) + (iPr)2NEtCl:HCl --> VX + HCl

SYNTHESIS

Step 1:

585 g (5 moles) of methylphosphonous chloride (SW) is added with stirring, over a 30 minute period to 690 g (15 moles) ethanol, in a nitrogen atmosphere, the temperature being kept at 25 °C. The mixture is allowed to stand for one hour at this temperature and then 160 g (15 moles) of sulfur and 1.5 liters of toluene is added and the system flushed with nitrogen. Ammonia gas is then passed rapidly through the vigorously stirred suspension maintained at 20-30 °C., until the exothermic reaction is complete (about 90 minutes). The mixture is then warmed to 50-60 °C. to expel any excess ammonia, then cooled, and stirred vigorously after addition of 1.5 liters of water. The aqueous layer is then separated, washed with 500 ml toluene, and if necessary, made neutral with a little conc. hydrochloric acid.

Step 2:

One mole of the ammonium salt of ethyl hydrogen methylphosphonothioate dissolved in 650 ml water is treated with 190 g (0.95 moles) diisopropylaminoethyl chloride hydrochloride and the mixture heated to 80 °C for 10 minutes with occasional stirring. The mixture is allowed to cool, and then extracted twice with 250 ml toluene, and the aqueous solution treated with 106 g (1 mole) sodium carbonate dissolved in 250 ml. water, and the VX extracted with toluene (750 ml, then 300 ml). The benzene is removed at reduced pressure (about 30 mm Hg), leaving a colorless residue.

B.P. #1,346,410 (1974), filed 1962, Wardrop and Stratford, U.K. Government.

Alternate VX Method #8

Step 3:   K + [MeP(S)(OEt)] + (iPr)2 >NEtCl --> VX

SYNTHESIS

Step 3:

50 ml of 0.4 M aqueous sodium hydroxide (of approximate pH 10) was added to approximately 100 ml of an aqueous solution containing about 9.17 g (0.04 moles) potassium ethyl methylphosphonothiolate.   Then   was   added   an   alkaline   solution,   pH   10,   of diisopropylaminoethylchloride. The solutions are allowed to stand for about five minutes. The pH of the solution is between 10 and 10.5, and after 5 minutes about 50 ml of .04 M acetic acid was added, and the pH dropped to about 5.15. The pH is adjusted and maintained at 10.5 followed by extraction with ether. The ether extract is separated and dried over anhydrous sodium sulfate, and the ether distilled off, leaving the VX.

Koblin, Abraham. U.S.P. #4,708,869 (1987), filed 1969, Dept. of the Army.

Conversion of Dimethylaminoethanol into Dimethaminoethylchloride Hydrochloride

Caution! This preparation should be conducted in a good fume hood.

In a 1 liter flask is placed 290 g (2.44 moles) thionyl chloride. The reaction flask must be cooled in an ice bath throughout the entire period of operation. 210 g (2.35 moles) of β-dimethylaminoethanol is added dropwise over a period of an hour, during which time there is copious evolution of sulfur dioxide. After the aminoethanol has been added, the ice bath is removed and the reaction mixture stirred for another hour. At this point the reaction mixture consists of a brown semisolid slush. The entire contents of the reaction flask are transferred to a 2 liter beaker, or flask containing approximately 1 liter of absolute ethanol. The resulting brown solution is heated to boiling on a hot plate, during which time there is a copious evolution of gases. The solution is filtered hot, leaving a small amount of insoluble material. Upon cooling of the filtrate on a salt-ice bath, the desired product is obtained as beautiful white crystals, which are collected on a Buchner funnel and dried in a vacuum desiccator over phosphorus pentoxide.

"Organic Syntheses, Coll. Vol. 4", p.333 (1963).

The Secret of Mr. Death: Homemade V-agent

"Twenty year for nothing, well that's nothing new,
Besides, nobody's interested in something you didn't do."

-- "Wheat Kings"
The Tragically Hip

For 35 years the secret held.

It was in a January 1977 "Playboy" magazine feature, an interview with a "Mr. Death", an alleged ex-weapons designer for the CIA.

> Mr. Death: Incidentally, in the [Devil's] Diary [manual] is an extremely simple
> method of synthesizing a rather potent nerve gas from a material
> that is easily available on your grocer's shelf right now. It requires
> no time or effort, really. ... It's not as toxic as VX...but damn close to it.

It was a V-agent, but not as toxic as VX. That was the clue.

Dimethylaminoethanol (DMAE) is available in the health food section of your average pharmacy or grocery store, and would make a V-agent with methyl groups instead of isopropyl groups at the end of the amine chain. DMAE is the chemical Mr. Death was referring to in the article, which could be used in place of diisopropylaminoethanol in any of the V-agent syntheses.

"Playboy" magazine, January 1977. Article: "Mr. Death"

Rothman, David B. "Mr. Death: The Life of a CIA Assassination Expert – By His Son". NY:
Playboy Press (1982)

BZ

BZ, or 3-quinuclidyl benzilate, is a hallucinogenic chemical warfare agent. It's not a nerve agent, but it was massively stockpiled by the U.S. as a chemical warfare agent, like one. The inhaled incapacitating dosage for BZ ranged from 125-215 mg min per cubic meter, or 0.5 mg orally.

While technically BZ is not an organophosphorus nerve agent, it **was** stockpiled by the ton as a non-lethal chemical warfare agent by the U.S.

Dr. MacArthur:

"BZ brings about complete mental disorientation, as well as sedation, which causes sleep. First of all the individual is completely confused as to what he is doing, or what he is supposed to be doing, and in addition he has hallucinations. He cannot carry out his assigned duties, nor can he remember what his assigned duties were. ... [Recovery] takes two or three days."

[Hearings, Subcommittee of the Committee on Appropriations, House of Representatives, Monday June 9, 1969]

BZ was originally synthesized by two Hoffmann-LaRoche chemists, Sternbach and Kaiser. The sodium salt of 3-quinuclidinol, from finely divided sodium metal and 3-quinuclidinol, was caused to react with diphenylchloroacetyl chloride, to form 3-quinuclidinyl diphenylchloroacetate. The latter was converted to the benzilate hydrochloride (BZ:HCl) by refluxing in 1N hydrochloric acid.

BZ: U.S.P. #3,919,241

U.S.P. #3,899,497

Piperidyl Glycollates: U.S.P. #3,903,094

Process for Synthesizing 3-Quinuclinyl Benzilate

SYNTHESIS

Into a 500 ml flask is placed 130 ml of dry toluene, to which is added 2.9 g (0.053 moles) of sodium methylate, 15 g (0.062 moles) methyl benzilate, and 5.2 g (0.041 moles) 3-quinuclidinol. The reaction mixture is slowly heated with stirring and about 65 ml of distillate containing the methanol formed during the reaction is distilled off.

The remaining solution is subsequently cooled to allow addition of 60 ml of toluene. The cooled reaction mixture also cooled to a temperature of 5 ºC, and 25 ml of water (with a temperature of 0-5 ºC) added. The solution is agitated until precipitation begins, whereupon it is stopped immediately, and the reaction mixture is cooled in an ice bath of one hour at a temperature of 0-5 ºC.

Upon completion of the one hour period, the mixture is suction filtered with a Buchner funnel, and the white filter cake is washed once with 10 ml of cold (0-5 ºC) toluene, then two washes of 5 ml each of cold (5 ºC) acetone.

Upon drying in air, the yield is 13.3 g BZ, which is a yield of 96.5% based on the 3-quinuclidinol.

U.S.P. #3,252,981 (1966), filed 1963.

U.S.P. #3,198,896.

U.S.P. #3,714,357 (1973), filed 1969.

Preparation of the BZ Precursor, 3-Quinuclidinol

Methyl isonicotinate was treated with ethyl bromoacetate, to yield the crystalline quarternary salt, which without further purification was hydrogenated to the diester. This diester was then cyclized with potassium metal. The keto-ester was saponified and decarboxylated in the usual way, and 3-quinuclidone isolated in about 40% yield in the form of its readily crystallizable hydrochloride salt.

3-Quinuclindinol was prepared from 3-quinuclidone by catalytic hydrogenation of its hydrochoride salt in the presence of platinum oxide. The basic alcohol forms white prisms or needles with the remarkably high melting point of 221-223 ºC.

Sternbach, L.H. " JACS" 74:2215-2218 (1952)

Acronymns

DC	methylphosphonic dichloride, MeP(O)Cl2
dichlor	DC
Et	ethyl (-C2H5 or -C2H4)
G-agent	RP(OR')(O)F
GB	sarin
GB	Great Britain Patent
iPr	isopropyl (-H(CH3)2)
iPrO	ispropyloxy (-OH(CH3)2)
JACS	Journal of the American Chemical Society
Me	methyl (-CH3)
OPA	iPrOH + isopropylamine
OR	alkoxy or aryloxy group (OMe, OEt, etc.)
R	alkyl or aryl group (Me, Et, Pr, iPr, cyclohexyl, phenyl, etc.)
sarin	GB (MeP(OiPr)(O)F)
SW	methyl phosphonous dichloride, MePCl2
TZ	isopropylamine
USP	United States Patent
U.K.	United Kingdom (Great Britain)
V-agent	R(R')P(O)(SEtN[OR"]2)

Bibliography

AD911 954. "The History of Chemical Warfare Plants and Facilities in the United States". ACDA/ST-197, Volume IV, U.S. Arms Control and Disarmament Agency, Midwest Research Institute (1972).

Curran, L.M. et al. "GB-DF Conversion Studies" Vol. 1 & 2, DRXTH-TE-CR-81109 & 81110, Battelle Columbus Laboratories, U.S. Army Toxic and Hazardous Materials Agency, Aberdeen Proving Ground, MD (1981).

"Chemical Disarmament: New Weapons for Old", SIPRI. NY: Humanities Press (1975). [Excellent!]

"Chemical Warfare Arms Control Inspection Handbook", ST-182, Midwest Research Institute, U.S.Arms Control and Disarmament Agency (1970?).

Hagihara et al. "Handbook of Organometallic Compounds". NY: W.A. Benjamin (1968).

Robinson, J.P.P. "The U.S. Binary Nerve Gas Programme", ISIO Monograph #10. Sussex, England (1975). [Excellent!]

Ward, F. Prescott. "Environmental Assessment ARCSL-EA-8101". "Construction and Operation of a 155mm M687 GB2 Binary Production Facility at Pine Bluff Arsenal, Jefferson County, Arkansas", Department of the Army (1981).

www.ingramcontent.com/pod-product-compliance
Lightning Source LLC
Chambersburg PA
CBHW080609180526
45168CB00007B/2836